The Professional Project Manager's Guide to Understanding Agile

in the

PMBOK® GUIDE
Sixth Edition

and the
Agile Practice Guide

The Professional Project Manager's Guide to Understanding Agile

in the

PMBOK® GUIDE
Sixth Edition

and the
Agile Practice Guide

JOHN G. STENBECK
PMP, PMI-ACP, CSM, CSP

Traditional. Agile. Hybrid.

First Edition | Spokane, WA

The Professional Project Manager's Guide to Understanding Agile
in *The PMBOK® Guide Sixth Edition*
and the
Agile Practice Guide

John G. Stenbeck, PMP, PMI-ACP, CSM, CSP

Published by:
GR8PM, Inc.
P.O. Box 0782, Deer Park, WA USA
(619) 890-5807

custserv@gr8pm.com; http://www.gr8pm.com/

ISBN 13 Edition: 978-0-9846693-8-7

This book includes references to terms such as PMP®, PgMP®, CAPM®, PMI-SP®, PMI-RMP®, or PMI-ACP®, all of which are registered marks of the Project Management Institute, Inc.

The Agile Practice Guide was published by the Project Management Institute and was developed jointly with the Agile Alliance.

The information contained in this book is provided without any express, statutory, or implied warranties. The author, GR8PM, Inc., its resellers, and distributors disclaim any liability for any damages caused or alleged to have been caused either directly or indirectly by this book.

Advance Praise

"John has done it again. He truly does have his finger on the pulse of project management and Agile. These two areas are on a collision course for a merger. John has focused on the highlights that will enable project managers and Agilistas to accelerate the merging of these two areas that are how projects and work gets done today. Thank you John for another must have book."

Wayne Brantley, MS Ed., PMP, PMI-ACP, CSM, ITIL, CPLP, CRP
Associate Vice President of Professional Education
Bisk Ventures

"Worried about Agile in the latest edition of PMBoK? Worry no more! This book explains why Agile was introduced and how it applies in today's Workplace. Project Managers can now arm themselves with the knowledge and language to talk to Engineers that are immersed in Agile in order to get information required for your stakeholders. A Must read for any Information Technology PMs out there!"

Deby Covey, MBA PMP CSM

"Thank you John for continuing to advocate for the beneficial intersection of agile and traditional project management practices. At the CIA, where I was employed for over thirty years as an engineer and project manager, I came to understand that my responsibilities included becoming proficient with many types of project life cycles and integrating them as needed to reflect the challenges of each particular project. One size did not fit all. That approach was reflected in the agency-wide project management program that I directed and in the first and only project management certification for US government civilian employees that I help develop and implement. This is precisely the type of integrative system thinking that you so wonderfully reflect in this book. The book content and the engaging writing style, elevate topics and concepts, that can be dogmatic elsewhere, to high levels of usefulness and practicability. Your book will be a big help to project managers looking to elevate their understanding of agile in the real world."

Michael O'Brochta, PMI-ACP, PMP
Author of "How To Get Executives To Act For Project Success"

"There are many seasoned professionals who have labored successfully in project vineyards using Agile methods and practices decades before the Agile Manifesto was written. John is one of them. This book reflects his depth of knowledge and understanding and will be a key resource for any novice or pro who wants help unpacking the new PMBoK Version 6. We are making it a "must read" for our clients and students."

Ted Kallman, FCT, PMP, PMI-ACP, CSM, CSPO, CSP
CEO, TIBA-USA

---

*"John has done it again! This is truly the Professional Project Manager's Guide to Understanding Agile in the **PMBOK® Guide, Sixth Edition** and the Agile Practice Guide. The analysis provided within these covers will save the Practitioner, the Academic Professor, and the Student hours of comparing and highlighting the Agile like enhancements to the **PMBOK® Guide** and the first edition of the Agile Practice Guide with other relevant publications such as the Agile Almanac Book 1 and Book 2.*

Discerning individuals will benefit from the 'Tinfoil Helmet Required Alerts', which clarify statements that could otherwise lead to more Agile 'Myths' and 'Anti-Patterns'. The Academic Professor will benefit from the clarity this book conveys in the Key Concepts, Tailoring Considerations, and the Considerations for Agile and Adaptive Environments sections. The Student and potential PMI-ACP Exam taker will become more informed by the attention getter, 'Exam takers should note' highlights.

Thanks for the numerous high-value 'Takeaways', which will most likely fuel countless spirited growth discussions among colleagues!"

Diane McCann, PhD, PMP, PMI-ACP
IT Integration manager, Brenntag North America, Inc.
Adjunct Professor, DeSales University and DeVry University

"Hands down – an outstanding and remarkably wonderful condensation of PMBOK and Agile principles. It gathers together all the related nuances in an elegant manner that is super easy to access, interpret and apply to add value immediately. It throttles down the golden nuggets in a truly seamless manner - allowing reader to focus on the essence (especially overlaps) instead of just the lexicons. John rightfully deserves our admiration for making this resource available."

Athar Safdar, MBA, PMP, PSM, PAL

"Thank you so much for saving the time it would take to get through another PMBOK edition! In this readable, distilled text, I got everything I need to know to continue teaching and evangelizing best practices in project management."

Taresa Nephew MBA, PMP
Program Manager at Itron, the largest IOT company in the world

"Finally, a highly credible guide that clears the air and provides an easy pathway on how to properly manage Agile projects in this world of mostly Waterfall projects! So excited to be a part of the GR8PM Tribe helping to lead, educate and influence others. Thank you, John, for your many years of leadership and expertise to help mature and refine agile project management!"

Debra L McDermott, PMP, CSM
TriWest IT PMO Director
Past PMI President, PMI Inland Empire Chapter 2009-2014

"Mr. Stenbeck has deep roots in the PMI community and is a project management evangelist at his core. AT first blush, I thought he was a bit harsh with some of his statements and opinions. After reading further, I see it differently and believe it was from the passion and concern he has for maturing our profession. There is NO ONE (or very few) who really understands the roots of Project Management as well he does, and NO One who has the un-biased opinion on where the future of our profession has been or is going. John speaks from a position of knowledge and experience with a clear vision of where we have come from and where we are going and translates it very well in this book. His Tin Foil Helmet alerts are deserving of your attention and consideration, not only if you are a new and upcoming project manager in search of your most important certification, but also to those who are long time credential holders. Reading this book has inspired me to take a deeper dive into the latest version of the PMBOK Guide, and I challenge you to do the same. There is a lot of new information, which if you are concerned about your career and keeping yourself relevant, whether or not your organization is, you need to know what the potential landscape of your future is going to look like! This is an important book you should read!"

John M Watson, PMP, PMI-ACP
PMI Northeast Fla Chapter
President 2008-2009
Director Emeritus 2010 – Present
Region 14 Component Mentor 2010 -2013

"Just finished reading the Professional PM's Guide to Understanding Agile in the PMBOK® Guide and APG, Joh Stenbeck's latest work. Really enjoyed the logical and pragmatic style, and was impressed with the depth of John's knowledge on Agile and Project Management. I had to go back and double check his credentials halfway through the book; I was convinced he must have a PhD that he didn't mention.

I got a kick out of the "Tinfoil Helmet Alerts" designed to help readers get over our usual double-take when the written word truly clashes with the real world of a project manager.

Also really liked the emphasis on "both/and" vs. "either/or" as related to Traditional and Agile approaches in project management. It's an important distinction that others seem to overlook, and I'm delighted to see a thought leader emphasizing the importance of this.

I appreciated the way he did the work of distilling what a typical project manager needs to know about the Agile additions. Sharing his insights and commentary on Agile practices along the way kept the book entertaining, insightful and relatable.

Thanks, John, for giving me the opportunity to review this exceptional work. In just a few short hours I was able to get a full grasp of what's included in the new Agile sections, and why it's important. The insights here saved me a ton of time in trying to interpret and relate my own experience to the subject. It is a true gift of time and insight!"

Susan Kennedy, PMP
Owner, American Realty
PMI Dallas Chapter, Past President 2018-2019; President 2016-2017; VP Professional Development 2011-2013

Table of Contents

Foreword

If you've been exposed to the *PMBOK Guide* before, either as a practitioner or as a student preparing for the PMP exam, you know that while the information it contains is incredibly valuable, it isn't always digestible. Practitioners and students now have to deal not only with Agile content being included in each knowledge area of the *PMBOK Guide Sixth Edition,* but also the content of the entirely new *Agile Practice Guide (APG).* Sometimes this material is merely obtuse while other times it is seemingly contradictory. The good news is that **The Professional Project Manager's Guide to Understanding Agile in the *PMBOK*® *Guide Sixth Edition* and the Agile Practice Guide** is available as a new "Rosetta Stone" helping make sense of all this new material as it works through seeming inconsistencies, and most importantly, translates these powerful new approaches into knowledge that you can start applying from day one.

In what I believe is his best and most readable work so far, John Stenbeck first delves into the Agile changes in the *PMBOK Guide Sixth Edition,* putting the always intimidating "Grid", new processes, renamed processes, and moved processes under the microscope. Even after having read the *PMBOK Guide Sixth Edition* myself, this book gave me many new insights that weren't immediately apparent. The five project lifecycles are covered in detail and special attention is given to key concepts such as the changes to Earned Value and Project Quality Management. You won't find any "Agile vs. Traditional" bias in this book, only knowledge that practitioners can tailor to help their Agile practices be more productive immediately. There are several memorable illustrations that will cement key concepts from the *PMBOK Guide Sixth Edition* in the mind of the reader.

The second section is devoted entirely to the all-new Agile Program Guide (APG) in a way that helps practitioners bridge the gap from Predictive to Agile approaches. In this section, a quantitative approach is taken cataloging and scoring elements of the APG to see how much value they provide. This is invaluable. Section Two clearly outlines what is considered in-scope and out-of-scope for the APG with scorecards provided for each. Section Two also goes into great detail helping the reader understand the nuanced differences between distributed and dispersed teams and tells you why you might get tripped up in this new language.

Whether your goal is to fully digest this new material and bring it to your workplace or to pass the PMI-ACP exam, the eminently readable and enjoyable *Professional Project Manager's Guide to Understanding Agile in the PMBOK® Guide Sixth Edition and the Agile Practice Guide* will help you customize and use it for your application.

Joe Montalbano, MBA, SPC, PMI-ACP, PMP, CSM, CSPO
Amazon #1 Best Selling Author, *Agile Almanac Book 2*

PART ONE:

The PMBOK® Guide – Sixth Edition

Introduction

If you've spent any time around PMI gatherings and PMP® credential holders you've heard the joke, "What is the definition of impossible?" and the answer, "Staying awake while reading the *PMBOK® Guide!*"

Given the non-readability of the *PMBOK® Guide*, we cast ourselves on your mercy as we try to make this guide as readable as possible, knowing that you understand it will never be a great novel that makes you want to sit down with a glass of fine wine, feet by the fire and dive into the next chapter. The trade-off will be the high-value career insights it will give you and the hours it will save you trying to read the *PMBOK® Guide* in order to stay professionally relevant.

With help from my Amazon #1 Best Selling co-author of the Agile Almanac Book 2, Doug Martin, PMP, PMI-ACP, CSM, CSP, we've undertaken this project with the goal of providing our Tribe *(currently over 18,000 members)* with an authoritative, comprehensive exploration and discussion that is also easily accessible, highly practical, and maybe even a bit humorous. We expect you will find great value here and, if you're not already part of the GR8PM tribe, perhaps a good reason to join us!

The release of the Project Management Institute, A *Guide to the Project Management Body of Knowledge, (PMBOK® Guide) - Sixth Edition, Project Management Institute, Inc. 2017* has created an interesting challenge for serious practitioners who really want to understand the meaning of the changes.

The challenge with understanding the *PMBOK® Guide - Sixth Edition* begins with the fact that it can be divided, at the highest level, into two core elements. Also, for PMI exam takers, it is coupled with the *Agile Practice Guide.*

- Part 1 – *The PMBOK® Guide - Sixth Edition* aligned to the Knowledge Areas with 13 chapters, including a reference section (541 pages).

- Part 2 – *The Standard for Project Management* aligned to the Process Groups (94 pages).

- The *Agile Practice Guide* with 6 chapters of Agile content independent of the *PMBOK® Guide - Sixth Edition* (138 pages).

(The *PMBOK® Guide* also includes various Appendices, a Glossary, and an Index (117 pages).)

The challenge emerges from the fact that even though *The Standard for Project Management* is part two, it underwent the revision process, as part of the *PMBOK® Guide,* <u>first.</u> That revision process does not appear to have taken into account the changes that would occur in the *PMBOK® Guide,* so part one and part two are not synchronized, potentially inducing misunderstandings and confusion.

The challenge grows when the independent content of the *Agile Practice Guide* gets factored into the equation. Written by an independent working group, which appears to have been more committed to providing Agile content as an "either or" choice instead of a "both and" choice even where core principles align, overlap and integrate but the choice of lexicon is different. It is akin to arguing over whether the pronunciation is "toe-may-toe" or "tah-mah-toe" instead of focusing on the value of the red fruit for making delicious sandwiches the customer loves (Figure 01). ***It is much ado about nothing!***

Figure 01

Despite the challenges, we congratulate PMI – and all of the contributors – on the decision to make a difference and move the state of our profession forward! We have a strong preference for prototypes, mock-ups, version 1, and iteration 0 to get something in front of the customer and initiate the unavoidable discussions that lead to high quality outcomes.

For the often-bureaucratic PMI this could signal an embrace of Agile practices, which could ignite the creative power of our worldwide membership. It was a stroke of leadership, regardless of whether it was by accident or by design, because it provides a baseline, a beginning, and a true call to action for PMI members to get involved with making Agile robust, useful, scalable, and fully integrated into the *PMBOK® Guide - Seventh Edition!*

The GR8PM Team is very pleased, and excited, about the potential that PMI has unleashed and we support it wholeheartedly!

Overview

Before deep-diving into the details it is important to **understand the magnitude** of the change that the *PMBOK® Guide - Sixth Edition* represents. Grasping the magnitude benefits from a brief review of the history behind the *PMBOK® Guide*. Recall that PMI was founded in 1969, and after a few versions that were more like whitepapers or collections of ideas, *The Project Management Body of Knowledge* was first published in 1987, almost 20 years later. Then in 1996, *A Guide to the Project Management Body of Knowledge* was first published, almost 10 years later. That publication was a watershed moment. The name change indicated a shift from the perspective that the body of knowledge was static and defined one that was dynamic and growing. The content, which added a ninth Knowledge Area – *Project Integration Management* – is the framework we all accept as gospel without realizing it was completely rewritten in 1996.

Imagining ourselves 10 years into the future, we can look back through the lens of time and see the *PMBOK® Guide - Sixth Edition* with better perspective. The name didn't change, but the framework was completely restructured. Agile was included in every Knowledge Area. It was released in English plus 11 other languages simultaneously, and coupled with the *Agile Practice Guide* raising the total page count of content from 616 pages to 936 pages, a *52 percent increase!*

Understanding the magnitude of the changes also means acknowledging that the changes were market-driven and critically important to the future of enterprise success – both public and private – across the globe, and fundamentally altered the expectations placed upon practitioners in the project management profession. Economies are no longer isolated to nations, regions, states or cities. The ubiquity of internet-enabled devices, social media, and big data have enabled the smallest of organizations to have global reach and the largest global organizations to be impacted by the actions of the smallest groups of interested stakeholders. That environment has made successful project management critical to enterprise survival and success.

Those market factors manifested themselves when Lean principles emerged with a new expression that found an almost feverish, dogmatic embrace within the software development industry proclaiming Agile as the panacea for all of the problems associated with project management and the *PMBOK® Guide.*

Despite a lot of hype, buzz and pronouncements to the contrary, the release of the *Agile Manifesto* in 2001 did not create the success of Agile, it merely benefitted from being in the right place, at the right time, when it expressed core principles that existed long before it. It benefitted from the perfect storm of supportive infrastructure, marketplace dynamics, and stakeholder engagement.

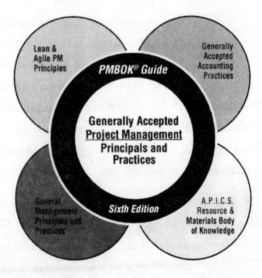

Figure 02

That same perfect storm is now demanding competencies and capabilities the *PMBOK® Guide* is better positioned to serve than the less mature, less robust Agile practices currently available elsewhere (Figure 02). The *PMBOK® Guide* has spent decades drawing upon the proven standards and best practices from Generally Accepted Accounting Principles (GAAP), the American Production and Inventory Control Society (APICS) body of knowledge, Lean principles, and general management principles and practices from business schools and universities.

The *PMBOK® Guide* is much more than just generally accepted project management principles and practices. It has now begun integrating leading Agile ideas because its mission and purpose is to be an authoritative curator of proven practices and information used in the project management profession. It is an ANSI Standard with the respect that is due such publications because of the rigor applied to its curation process and the contributions invested by many hundreds of professional practitioners. It carries that responsibility with sober discipline and has earned immense marketplace credibility because of it.

In the 30 years since the original, *The Project Management Body of Knowledge*, was published, the *PMBOK® Guide* has continued to assess, analyze, codify and integrate validated proven concepts from a host of sources into **the standard** that is relied upon by every major enterprise using, or wishing to use, project management to improve results by delivering on constituent, customer and stakeholder expectations.

The fact that the *PMBOK® Guide* is *the standard* is conveyed in a concise and instructive comment, highlighted perhaps by its understatement, in Chapter 1, Section 1.2.4.1, Project and Development Life Cycles. It says, "A project life cycle is the series of phases that a project passes through from its start to its completion. It provides the basic framework for managing the project. This basic framework applies regardless of the specific project work involved. The phases may be sequential, iterative, or overlapping." It is a simple yet profound fact that all projects can be mapped to a generic life cycle at the highest level (Figure 03).

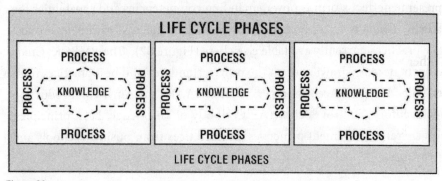

Figure 03

The generic life cycle can be understood as knowledge being created and used, and developing within a process, that is integrated as part of a life cycle.

The statement in Chapter 1, Section 1.2.4.1 continues, "Project life cycles can be predictive or adaptive. Within a project life cycle, there are generally one or more phases that are associated with the development of the product, service, or result. These are called a development life cycle. Development life cycles can be predictive, iterative, incremental, adaptive, or a hybrid model," as shown in Figure 04.

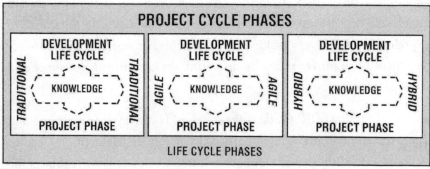

Figure 04

It also says, "It is up to the project management team to determine the best life cycle for each project. The project life cycle needs to be flexible enough to deal with the variety of factors included in the project."

Finally, it states, "*Project* life cycles are independent of *product* life cycle, which may be produced by a project. A product life cycle is the series of phases that represent the evolution of a product, from concept through delivery, growth, maturity, and to retirement." *(Emphasis added.)*

In other words:

- *Predictive* life cycles set the project scope, time, and cost early in the life cycle, and then manage changes carefully. Predictive is also known as (AKA) waterfall or traditional life cycles.

- *Iterative* life cycles also set the project scope early, as a rule, but time and cost are expected to be modified as the project team gains a better understanding of the product. The product develops through repeated cycles, called iterations, delivering increments that add functionality to the product.

- *Incremental* life cycles add functional deliverables in iterations within an agreed upon timeframe, but contain the complete, sufficient capabilities only after the final iteration.

- *Adaptive* life cycles start with a detailed scope and use iterations to develop the product and can be iterative or incremental. AKA Agile or change-driven life cycles.

- *Hybrid* life cycles combine predictive and adaptive, or multiple adaptive frameworks, to create a life cycle tailored to the needs of the team and environment.

Quick Review of the *Agile-Specific* Revisions

At a more granular level the *PMBOK® Guide*, Chapter 1, also includes the Project Management Process Group and Knowledge Area map, affectionately known as, *"The Grid!"* by PMP® credential holders the world over. The five Process Groups have remained the same, however, the Grid has two renamed Knowledge Areas, three new Processes, nine renamed Processes, and one relocated Process.

Project Time Management and Project Human Resource Management were renamed as the Project Schedule Management and Project Resource Management Knowledge Areas, respectively, emphasizing the critical nature of scheduling and the broader responsibility for both team and physical resources.

Three new processes have been added. First, Manage Project Knowledge was added to the Executing Process Group where it intersects the Project Integration Management Knowledge Area. Second, Implement Risk Responses was also added to the Executing Process Group where it intersects the Project Risk Management Knowledge Area. Third, Control Resources was added to the Monitoring and Controlling Process Group where it intersects the Project Resource Management Knowledge Area.

Nine processes have been renamed. Perform Quality Assurance is now Manage Quality. Plan Human Resource Management is now simply Plan Resource Management. Acquire Project Team has become Acquire Resources. Develop Project Team and Manage Project Team both dropped "Project" to become simply Develop Team and Manage Team. Control Communications is now Monitor Communications, and Controls Risks is likewise Monitor Risks. Lastly, Plan Stakeholder Management and Control Stakeholder Engagement has become Plan Stakeholder Engagement and Monitor Stakeholder Engagement. The processes were renamed to clarify the alignment with the process intent and aid understanding.

Estimate Activity Resources remains in the Planning Process Group, but was moved and now intersects the Project Resource Management Knowledge Area.

The processes have been categorized into three groups – processes that occur once or at predefined points, periodic processes done as needed, and repeated or continuous processes.

Deep Dive of the Agile Revisions

This deep dive is specifically, and exclusively, focused on the revisions to the PMBOK® Guide - Sixth Edition that relate to the Agile content that was added and integrated. There are, of course, many other revisions related to the non-Agile sections of the *PMBOK® Guide* and an army of PMI Registered Education Providers can provide coverage of those to help practitioners integrate them.

Any sections that do not contain changes we considered important to Agile practitioners and PMI-ACP® exam takers have been left out with the expectation that doing so will optimize the time required to fully grasp the Agile changes. Being practitioners ourselves, we understand what a precious commodity time is, how short the supply of time is for project managers, and therefore have made this decision to better serve them.

CHAPTER 1

Introduction

Section 1.2 Foundational Elements

As explained above, section 1.2.4.1 defines project life cycles, product life cycles, and predictive, iterative, incremental, adaptive, and hybrid models. It also says:

- The project management team should use care to select the best life cycle.

- The chosen life cycle must have the flexibility to accommodate all the variables that will impact the project.

- Project life cycles should be understood as separate from product life cycles, but that integrating them is important in order to deliver and evolve the product successfully from concept to maturity and retirement.

Section 1.2.4.4 has detailed coverage of the three process types. It says:

- Examples of processes used once or at predefined points include *Develop Project Charter* and *Close Project or Phase*.

- Examples of processes performed periodically include *Acquire Resources* and *Conduct Procurements*.

- An example of a process performed continuously throughout the project is *Define Activities.*

CHAPTER 2
The Environment in which Projects Operate

Chapter 2 includes information on Enterprise Environmental Factors (EEFs), Organizational Process Assets (OPAs), processes, policies, and procedures (Section 2.3.1), however no significant Agile-related changes were included. That, of course, struck us as odd since it is the environment in which projects now operate that drove the inclusion of Agile in the *PMBOK® Guide.*

As we noted in the introduction, the *PMBOK® Guide – Sixth Edition* is an excellent first iteration, meaning it is also an invitation to add great value and content to the *Seventh Edition* by volunteering to be part of that PMI committee. This is one of the many opportunities for serious practitioners, such as you, since you are reading this, to get involved as an active PMI volunteer and help us build the amazing future that lies before our wonderful profession!

CHAPTER 3

The Role of the Project Manager

The *PMI Talent Triangle®* is highlighted in Chapter 3 as part of the coverage of *The Role of the Project Manager*. This chapter discusses organizational responsibilities and the skills and competencies required to be effective project managers. It expands coverage from 2 pages in the *PMBOK Guide® - Fifth Edition* to 17 pages in the new edition.

Even though it discusses leadership without any explicitly Agile perspectives, it is fundamentally important because Agile succeeds only in the presence of real leadership. Therefore, we will address it briefly.

The *PMI Talent Triangle®* defines three key facets as:

- *Technical project management* highlighting the knowledge, skills, and behaviors specific to the project, program, and portfolio management domain required to properly perform required duties.

- *Leadership* highlighting the knowledge, skills, and behaviors needed to help teams achieve business and organizational goals.

- *Strategic and business management* highlighting the industry and organizational knowledge and expertise required to enhance performance and better deliver business outcomes.

Section 3.5, Performing Integration, says the role of the project manager is two-fold:

- First, project managers must work effectively with the project sponsor to align project objectives with strategic goals and ensure timely, desirable results that contribute to successful strategy execution.

- Second, project managers help teams succeed by guiding the integration of processes, knowledge, and team efforts so they stay focused on essential project outcomes.

It also notes that integration takes place at the process, content and context levels, as follows:

- *3.5.1 Performing Integration at the Process Level.* If the project manager is unsuccessful integrating project processes poor, even catastrophic, results become a major risk. The challenge is that integrating project processes requires sophistication and skill that cannot be defined as a process itself.

- *3.5.2 Integration at the Cognitive Level.* Specific characteristics, such as project size, uncertainty, complexity and regulatory or organizational standards, drive the sophistication required to integrate the processes. Therefore, the cognitive abilities and skills of the project manager play a critical and unavoidable role in whether it is possible to achieve the desired project results.

- *3.5.3 Integration at the Context Level.* Context is often overlooked as a factor in project management even though it exerts a vital influence on stakeholder expectations. The impact of contextual variables, such as communication planning, knowledge management, and the need for learning and discovery, play a major role in successfully guiding the project team. Project managers must be cognizant of the power of context when managing integration.

Chapters 4 through 13 are where the new Agile content is most explicit and the changes to the structure the most significant. Knowledge Areas in prior editions of the *PMBOK® Guide* were expressed using an introduction, outline, and overview illustration followed by the content. In the Sixth Edition, the structure of each Knowledge Area has four elements, *Key Concepts, Trends and Emerging Practices, Tailoring Considerations, and Approaches in Agile, Iterative, and Adaptive Environments.*

Placing the three sections, *Trends and Emerging Practices, Tailoring Considerations*, and *Approaches in Agile, Iterative, and Adaptive Environments* at the beginning of each Knowledge Area would seem to imply the importance of the changes being driven by Agile principles and practices.

Key Concepts covers the information in the first three sections of prior editions. *Trends and Emerging Practices* expresses what is considered a good practice for a majority of projects, most of the time, and also includes a limited discussion of industry trends not practiced on most projects, explaining why they are part of the process inputs, tools, techniques, and outputs (ITTOs). *Tailoring Considerations* highlights options for tailoring the processes, ITTOs, and lifecycles, and includes a list of questions useful to practitioners as they customize their project management approach. *Approaches in Agile, Iterative, and Adaptive Environments* covers development methods, techniques, artifacts, and practices specific to the use of Agile, iterative, and adaptive approaches. Because Agile techniques have been integrated throughout the Sixth Edition, specific ideas to help practitioners identify and integrate them into their projects is also provided.

CHAPTER 4
Project Integration Management

Under *Key Concepts* for Project Integration Management the first thing pointed out is that this Knowledge Area is specifically for project managers, suggesting it does not involve other stakeholders like the remaining Knowledge Areas do. Many of the other Knowledge Areas involve and engage team members, such as schedulers and cost and risk analysts, but responsibility for project integration cannot be delegated. As noted above, integrating project processes requires sophistication and skill that the project manager must possess and apply while combining interim results for an overall view of the project (Figure 05). That view includes:

Figure 05

- Ensuring due dates for product deliverable are met.
- Planning so that project objectives are achieved.
- Supporting knowledge creation and appropriate use of discoveries.
- Managing performance and adapting to changes.
- Applying integrated decision-making to change management.
- Communicating information effectively to stakeholders.
- Managing transitions while balancing competing considerations.

The project manager must keep in mind that the more complexity and uncertainty affecting the project, the more varied stakeholder expectations will be, thereby demanding an increasingly sophisticated approach to integration management.

In *Trends and Emerging Practices* four key concepts highlighted are:

- Visual management tools
- Project knowledge management
- Project manager responsibilities
- Hybrid methods

Figure 06

Whether the enterprise is public (taxpayer funded) or private (consumer funded) complexity has been driven by stakeholder expectations altered by Google, Amazon, and Facebook (Figure 06). Professional PMs must improve communication by deftly applying visual management tools, such as burn-down charts, leverage knowledge management to include learning and discovery, and accept accountability for demonstrating sophisticated integration skills, which includes tailoring approaches to engage hybrid methods best suited to meet altered stakeholder expectations.

Tailoring Considerations emphasizes each project's unique context and content, defining the project manager's need to tailor project integration management by listing factors for tailoring that include:

- Project life cycle
- Development life cycle
- Management approaches
- Knowledge management

In the section on *Considerations for Agile and Adaptive Environments* the expectation that the project manager can effectively delegate control of detailed planning and delivery to the team in a collaborative, decision-making environment is identified as critical to success in adaptive environments.

Stakeholder expectations of the PM do not change just because the project is in an adaptive environment. While the PM may delegate control of the detailed planning and delivery of increments of the solution to the team, the PM remains accountable for collaborative decision-making that includes stakeholder feedback in order to respond effectively to changes.

We often tell our clients and students that there are times when the "real world" and the world of the *PMBOK Guide®* don't align, making it necessary, especially for exam takers, to don the proverbial "tinfoil helmet" (to prevent "aliens" from attacking your brain). There are also equivalent times when Agile-speak or Agile dogma induces the need for a tinfoil helmet. When those occasions arise, we will use the "ALERT – TINFOIL HELMET REQUIRED" icon to draw your attention to them.

ALERT

TINFOIL HELMET
REQUIRED

The first example of a Tinfoil Helmet alert occurs here in Chapter 4, Tailoring Considerations, when the *PMBOK® Guide* says collaboration is enhanced when, "… team members possess a broad skill base rather than a narrow specialization." Real world experience is more nuanced. The <u>team</u> needs a broad skill set, which is why it must be cross-functional, but that requires that each team <u>member</u> not provide a specialized skill set.

While the concept, often called "generalizing specialists", suggests the hypothesis that the type of person able to master multiple skills has a strong ability to learn new skills quickly, the idea seems to have emanated from the software industry referring to programmers who also do other development-related tasks, such as testing. Yet, within software development this concept has limited acceptance and outside of software development it has virtually none.

Generalizing specialists sounds like a good idea but is a misconception in the sense of being a less-than-robust response to the central challenge (Figure 07). In this case, the central challenge is how to avoid the risk of dependency on

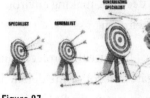

Figure 07

sole-source providers of particular skills because, among other things, they can limit the ability of the project to make progress or they can leave the organization, creating a major operational deficit. Let's look at this concept from different viewpoints.

From an accounting perspective, each member of the team is a resource paid in proportion to the tacit knowledge they supply. An increase in <u>tacit</u> knowledge, also known as experience or battle scars, is why a typical engineer with 10 years of experience gets paid more than a recent college graduate in the same field. If the more experienced engineer leaves an organization, it is likely harder, and more expensive, to fill the void.

Figure 08

An alternate, and perhaps stronger, solution for this problem is <u>tribal</u> knowledge. Consider that the Olympic track and field relay team with the fastest runners doesn't always win the race. But without fail, 100 percent of the time, the winning Olympic relay team is one that did not drop the baton. The baton is the symbol of tribal knowledge and the runners are the sole-source providers of skill (Figure 08).

One of the powerful factors common to Agile approaches is cross-functional teams. Each member of the cross-functional team contributes tacit knowledge while the system of Iteration Planning, Stand-up Meetings, User Stories and Planning Poker creates tribal knowledge. This is available because the system strengthens the interface, the hand-off of the baton, with continuous process improvement. The system creates tribal knowledge, reducing development risks while also mitigating the impact of losing a team because it reduces the transition cost and learning curve for the new hire when an existing team member explains how the interface works so the baton doesn't get dropped.

This approach also compensates for a nearly-fatal flaw in the assumptions underpinning the "generalizing specialists" solution. In a great many envi-

Figure 09

ronments, sole-source skill providers are as different as zebras and giraffes. Asking them to be generalizing specialists is like hoping for a zebra-striped giraffe (Figure 09). Agile offers a much more reliable solution with a systemic approach, creating easy-to-hold batons of tribal knowledge.

CHAPTER 5

Project Scope Management

Under *Key Concepts* for Project Scope Management, without specifically saying it is a difference commonly overlooked by project managers, it says the PM needs to differentiate two related but entirely different ideas (Figure 10). They are:

- *Product Scope* which is the features, functions and capabilities (FFC) of the desired outcome; the tangible or intangible product that is also sometimes a service.

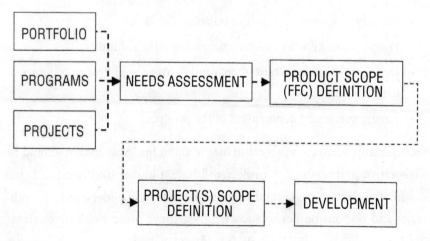

Figure 10

- *Project Scope* which is the work that must be successfully completed to deliver the product scope. Project scope can exceed product scope, such as including a marketing campaign to accompany the product. It can also be less than the product scope, such as being a Phase 1 Release, that is, a subset, of the final product.

The difference has many important impacts, such as measuring the project scope against the plan versus measuring the product scope against the product requirements.

In *Trends and Emerging Practices*, it points out that requirements have always been important and become even more critical as the global environment becomes more demanding. As noted earlier, the evolution of stakeholder expectations, driven by their experience of Google, Amazon and Facebook, have made the complexity of organizational environments rise exponentially. It is, therefore, vital to recognize that the use of business analysis directly impacts competitive advantage whether the enterprise is pursuing taxpayer or consumer funding.

Business analysis activities may start before a project is initiated in order to:

- Identify problems and resolve business needs.
- Frame viable choices and recommend specific solutions.
- Define stakeholder requirements and manage workflow in order to meet objectives
- Ensure successful deployment of the product

The relationship between a project manager and a business analyst should be a collaborative partnership. A project will have a higher likelihood of being successful if project managers and business analysts fully understand each other's roles and responsibilities to successfully achieve project objectives. Their disciplines are different, but their success is intertwined.

The *Tailoring Considerations* section lists five areas:

- Knowledge and requirements management
- Validation and control
- Development approach
- Stability of requirements
- Governance

In the section on *Considerations for Agile and Adaptive Environments* it says on projects with high complexity, uncertainty, or risk, significant evolution of the scope is unavoidable as learning and discovery by the team is shared with the stakeholders and their understanding of their needs is clarified (Figure 11). During the project, Agile methods deliberately spend more time in the process of discovery and refinement as a source of risk mitigation caused by the gap between real requirements and those originally stated. In Agile approaches, requirements are developed and managed using a logical device for information management referred to as a Backlog.

Figure 11

CHAPTER 6

Project Schedule Management

For Chapter 6, the initial ideas in **Key Concepts** include:

- Schedules are a detailed representation of how delivery of increments, and final outcomes, are expected to occur.

- Schedules are a communication tool for managing stakeholder expectations.

- Schedules provide a basis for performance reporting.

Notice that the *PMBOK® Guide* says schedules <u>represent</u> delivery, conveying an acceptance that they will change as reality unfolds, with all of its twists and

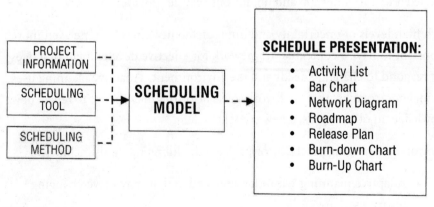

Figure 12

turns, during execution. Taking into account project information, the capabilities of the scheduling tool, and the choice of scheduling method, the project manager and team develop a scheduling model that generates the schedule presentation (Figure 12).

That means schedules clearly benefit from retaining flexibility so they can integrate insights from knowledge gained, risks clarified and value-added activities identified.

In *Trends and Emerging Practices* it says that the global marketplace where enterprises compete has made it vital to have a contextual framework for adapting and tailoring development methods because long term scope is so difficult to define as new knowledge emerges. New knowledge drives changing priorities, which is the essence of a new context.

Figure 13

Repurposing the popular adage that, "Culture eats strategy for breakfast" our experience validates that, "Context circumscribes content." That means successful enterprises must get the context right and integrate it with the right content (i.e., technology) as disruptors, like Uber for instance, emerge (Figure 13). Uber got the context of the rider-driver relationship correct and married it to a friendly app and ubiquitous service providers.

With high levels of uncertainty and unpredictability it has become even more important to have a contextual framework for effective development practices that respond to changing needs and the environment. Adaptive planning uses a plan, but recognizes that once work starts priorities may change to reflect new knowledge, in other words, a new context.

Some of the emerging practices for project scheduling methods include:

- Adaptive planning with new knowledge that may drive changing priorities.

- Agile product development using iterative schedules and incremental deliverables.

- On-demand scheduling, like Kanban, applying concepts from Lean manufacturing to increase throughput via continuous flow production.

The *Tailoring Considerations* section lists four considerations:

- Lifecycle approach – Which is most useful?

- Resource availability – Which factors should influence choices?

- Project dimensions – Which elements intersect with control needs?

- Technology support – Are the right technologies available?

Each is relatively self-explanatory. The important point is that the decisions are made by design and not by accident.

In the section on *Considerations for Agile and Adaptive Environments* it aptly notes that in large organizations an ever-changing mix of small projects and large initiatives coexist. Therefore, the use of long-term tools, such as Roadmaps or Program Plans, to manage scaling factors, such as varying team sizes, divergent geographical distribution, changing regulatory requirements, organizational flux and technical complexity, impacts the PM and the team. To effectively cope with these challenges organizations often need to combine practices, or adopt a method that has already done so, to create a manageable environment.

Such environmental factors don't change the role of the project manager, even when projects are managed in an adaptive way. To be successful the project manager will need to be familiar with all the tools and techniques – Traditional and Agile – and understand how to apply them.

CHAPTER 7
Project Cost Management

Under *Key Concepts* three highlighted ideas are:

- Cost management is primarily focused on the cost-to-complete project activities.

- Cost management should consider the product's Total Cost of Ownership (TCO) when making project decisions.

- Project costs can be capital investment or operational expenses, and the financial impact of mishandling or classifying them incorrectly can have a grave bearing on perceived project success and PM competence.

Another idea that is critical to successful PMs is that different stakeholders apply different metrics when assessing success, and those metrics may vary over time. Regular, effective communication is the only antidote to the risk of changing metrics and requirements.

In *Trends and Emerging Practices* a very interesting emphasis, directly impacting the credibility of scaling Agile, is covered. The expansion of Earned Value Management (EVM) to include Earned Schedule (ES) addresses the need for reliable, generally accepted progress reporting.

Traditionally, the most used EVM formulas have been Schedule Variance and Index and Cost Variance and Index (Figure 14).

Schedule	Variance	=	EV	–	PV
Schedule	Index	=	EV	/	PV
Cost	Variance	=	EV	–	AC
Cost	Index	=	EV	/	AC

Figure 14

Earned Schedule theory expands the metrics used in traditional EVM with two new variables, Earned Schedule (ES) and Actual Time (AT) (Figure 15). Using the equation ES – AT, if the amount of ES is greater than zero, then the project is ahead of schedule by the amount shown. In other words, Schedule Variance quantifies how much more *progress than planned* the project earned at a given point in time. Using the equation ES / AT, if the Schedule Index is greater than one, it indicates the *efficiency of development* work and the return on investment (ROI) from project resources is positive by a specific rate at a given moment in time.

Schedule	Variance	=	ES	–	AT
Schedule	Index	=	ES	/	AT
Cost	Variance	=	EV	–	AC
Cost	Index	=	EV	/	AC

Figure 15

These capabilities of ES could significantly improve practitioner's ability to quantify and substantiate reporting of Benefits Realization thereby adding significant motivation to use Agile!

The *PMBOK® Guide* says, "ES is an extension to the theory and practice of EVM. Earned schedule theory *replaces* the schedule variance measures..." *(Emphasis added.)* We think serious practitioners should use it to augment, not replace, the older measures because, in our experience, quantifying schedule variance in monetary units is a very good way to capture the often limited, hard-to-reach attention of key stakeholders.

The section on *Tailoring Considerations* stresses five points that include:

- Knowledge management – Consider the cost versus value of knowledge, and communicate it effectively.

- Estimating and budgeting – Use the concept of "applied granularity" to help stakeholders properly respond to the information.

- Earned Value Management – As discussed above, use EVM to help with reporting benefits realization.

- Agile approaches – Offer methods for integrating the first three points above.

- Governance – Guides aligning with regulatory and audit requirements and can facilitate the implementation of scaled agile methods.

In the *Considerations for Agile and Adaptive Environments* section it spotlights two thoughts:

- Projects with high-uncertainty or undefined scope may not benefit from detailed cost calculations because of the high probability that scope will change as learning and discovery occur.

- Lightweight estimation methods can be much more cost effective, generate high-level forecasts of project costs to appropriately support decision-making faster, and be adjusted as stakeholder approved changes arise.

CHAPTER 8
Project Quality Management

The **Key Concepts** discussion points out that it involves both the projects and deliverables. It starts with three points, as follows:

- Failure to meet quality requirements can have serious, negative consequences for stakeholders.

- *Quality* and *grade* are not the same. Quality measures requirements fulfillment. Grade measures the alignment of functional use against different characteristics.

- The PM and team must carefully manage trade-offs between quality and grade (Figure 16).

Figure 16

It then states that prevention is preferred to inspection, which completely aligns with Lean and Agile principles. It notes:

- Prevention keeps errors out of the process. Inspection keeps errors out of the deliverables.

- Attribute sampling measures conformance dichotomously while variability sampling measures conformation across a continuum.

- Tolerances specify acceptable ranges while control limits define boundaries for statistically stable performance.

Successful PMs must manage both the cost of quality (COQ) as well as the cost of poor quality or failure costs.

In *Trends and Emerging Practices* highlights the core idea that quality management seeks to minimize variation and deliver results that meet stakeholder requirements. Considerations that are identified include:

- *Customer satisfaction* which is defined as a combination of conformance to requirements and fitness for use that satisfies real needs.

- *Continual improvement* is based upon the Plan-Do-Check-Act (PDCA) cycle defined by Shewhart and modified by Deming. In addition, quality improvement initiatives such as total quality management (TQM), Six Sigma, and Lean Six Sigma are noted as ways to improve the quality of project management, the end product, and overall results.

- *Management responsibility* is specified as the responsibility to provide adequate resources with suitable capabilities to engage the team for a successful outcome.

- *Mutually beneficial relationships with suppliers* are interdependent, collaborative and demonstrate a preference for longer-term, overall advantage versus shorter-term gains because the longer-term focus contributes to a continuously improving value stream for both parties.

The *Tailoring Considerations* section calls out four points:

- *Policy compliance and auditing* with an orientation toward creating policies and procedures supported by tools, techniques and templates.

- *Standards and regulatory compliance* using the most applicable standards and best practices.

- *Continuous improvement* with a clear understanding of whether it will be driven at the organizational level or managed at the project level and how they are interrelated.

- *Stakeholder engagement* enabled through collaborative standards and transparent communication.

The *Considerations for Agile and Adaptive Environments* section starts with the observation that recurring retrospectives create insight into the effectiveness of the quality processes. They also provide root cause analysis during team discussions that identify ideas to overcome the obstacle or issue.

The next statement, while true, risks creating stakeholder confusion. It says, "… agile methods focus on small batches of work, incorporating as many elements of project deliverables as possible." The confusion occurs because of the contrast of "small batches" and "as many elements as possible." Recognize that small batches govern how many elements can be included.

ALERT

TINFOIL HELMET
REQUIRED

Section 8.1.2.3, Data Analysis includes a list of *cost of quality* factors a project should consider. It is important to weigh the cost of conformance against the cost of non-conformance (Figure 17). They include:

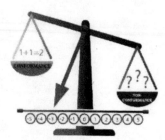

Figure 17

- *Prevention costs,* defined as prevention costs to protect products, deliverables, or services of the project from outcomes with poor quality.

41

- *Appraisal costs,* defined as the cost of evaluating, measuring, auditing, and testing project deliverables.

- *Failure costs,* which can be internal or external and relate to non-conformance to stakeholder needs or expectations.

The optimal cost of quality occurs at the point where the cost of prevention and the cost of future failures are in balance or equilibrium. This should be kept in mind so that investing in prevention or appraisal costs that are not beneficial or cost effective does not occur.

CHAPTER 9
Project Resource Management

Key Concepts defines the project team as individuals working together to achieve a shared goal. It follows, therefore, that involving all team members in project planning and decision-making is beneficial because it leverages their expertise and catalyzes their commitment.

Successful PMs must embrace both leadership and management responsibilities. In addition to managing *things,* such as schedules, work execution and stakeholder communication, they must lead *people* by influencing things such as team formation, skill development, motivation and team satisfaction.

It is also crucial that they not only model high ethical standards, but also enforce them (Figure 18).

Figure 18

Trends and Emerging Practices begins with saying, "Project management styles are shifting away from a command and control structure for managing projects and toward a more collaborative and supportive management approach that empowers teams by delegating decision-making to the team members."

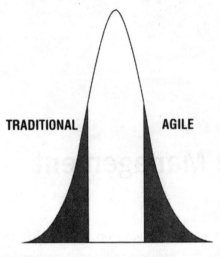

TRADITIONAL **AGILE**

Figure 19

The point being made may be true at the micro-dynamic, team level of project management, but it misses two important conditions. First, in our experience, in large enterprises about 30 percent of projects benefit from the added overhead of using Agile methods. There are very few large organizations where more than 30 percent of the all projects are using an Agile approach. Second, it is not true at the program, portfolio or enterprise macro-dynamic level, except for Kanban, which is acquiring more and more traction as a program and portfolio management approach. Accepting the assertion that "styles are shifting away from command and control" requires an acceptance that the distribution of projects is a very tall, narrow bell curve, and that Agile has crossed the chasm and moved into the mainstream in industries outside of software development (Figure 19).

The section continues by identifying four ways modern project management seeks to optimize resource utilization. The concepts include:

- *Resource management methods* such as Lean management, theory of constraints and total productive maintenance.

- *Emotional intelligence (EI)* being applied to help teams become more emotionally competent and reduce employee attrition.

- *Self-organizing teams* with the support and trust needed to deliver successfully.

- *Virtual and distributed teams* driven by the globalization of development and production approaches.

Two alerts are needed here. The first alert is that the statement about self-organizing teams says, "... where teams func-
tion with an *absence of centralized control*." *(Emphasis added.)*
Such a statement may be 'incendiary' for stakeholders who
will imagine a state of chaos occurring. It would be better
explained as, "Where the team exercises self-determination
within the boundaries or scope of the Iteration."

ALERT

TINFOIL HELMET
REQUIRED

The second alert is about the virtual and distributed teams discussion. It says,
'...globalization has created the need for virtual teams that are not colocated'
and then lists examples of challenges. The list is missing the biggest challenge,
a common language. For the vast majority of virtual teams in the real world,
English-as-a-Second-Language (ESL) is the biggest challenge.

Tailoring Considerations highlights six ideas by asking a question of each,
including:

- Diversity - "What is the team's background?"

- Physical location - "Where are the team and physical resources
  located?"

- Industry specific resources - "What special resources are needed?"

- Acquisition of team members - "How are team members acquired,
  and are they full- or part-time?"

- Management of team - "How will team capabilities be managed
  and improved?"

- Lifecycle approaches - "Which one will the project use?"

Considerations for Agile and Adaptive Environments makes the important
point that collaboration is intended to boost productivity and facilitate innova-
tive problem solving. It also correctly points out that projects with high-variability – ones with high complexity and high uncertainty – benefit from team
structures that maximize focus and collaboration.

45

Unfortunately, it then goes on to repeat the Agile dogmatic canard about 'self-organizing teams with generalizing specialists' that we debunked earlier (Figure 20).

Figure 20

We think clarity would be enhanced by using a lexicon aligned with the decision-making levels where:

- Self-organizing teams are understood to operate with autonomy at the Iteration and Milestone decision-making level.

- Self-directing Teams are understood to operate with autonomy at the Release and Project decision-making level.

- Self-managing Teams are understood to operate with autonomy at the Roadmaps and Program/Portfolio decision-making level.

CHAPTER 10
Project Communications Management

Project communication management *Key Concepts* focuses on the exchange of information, by accident or design, using a variety of elements, such as:

- *Written form,* which can be analog or digital.
- *Spoken,* which can be in person or electronically mediated (Figure 21).
- *Formal or informal,* which can be policies or blog posts.
- *With context,* which includes gestures, tone, volume, facial expressions, and word choices.

Figure 21

- *Through media,* which includes dropdown menus, flowcharts, mind maps, pictures, and videos.

Communication, internal and external, consumes most of a successful PM's time because building relationship bridges, or failing to do so, reduces or causes risk. PM success comes from a solid strategy and effective tactical implementation.

Communications management in *Trends and Emerging Practices* makes clear that developing and implementing appropriate communication strategies is vital to maintaining effective relationships with stakeholders. Practices highlighted include:

- Include everyone identified as essential to successful project outcomes in project reviews.

- Understand how users engage social computing to build relationships and network with them.

- Respect cultural, practical, and personal preferences for language, media, content, and delivery.

Tailoring Considerations outlines these five elements to integrate in decision-making:

- Stakeholders
- Physical location
- Communication technology
- Language
- Knowledge management

Considerations for Agile and Adaptive Environments explores the fact that projects have varying degrees of ambiguity and change where discovery and learning will drive the frequency, intensity and duration of communications needed to be effective.

Having artifacts that are transparent and easily accessible for the stakeholders to review will increase the probability of success in these environments.

CHAPTER 11
Project Risk Management

Key Concepts begins with a fundamental statement of truth. All projects are risky, with varying degrees of complexity, in a context of constraints and assumptions responding to stakeholder expectations that may conflict or change. It is a mouth full, but accurate and complete.

Two levels of risk exist as individual risk and the overall riskiness of the project. Risk arises from the combination of individual risks and other sources of uncertainty.

- *Individual project risk* at the Iteration and Milestone level.

- *Overall project risk* at the Roadmap and Program level.

Risks are defined as being both positive and negative depending on their effect on project objectives, if they occur. Risks continue to emerge throughout the project so processes should be conducted iteratively. It is also important to have key stakeholders decide on thresholds regarding what level of risk exposure is acceptable. Acceptability needs to be defined with measurable thresholds that reflect the risk appetite of the organization and the project stakeholders.

The understanding of risk management in *Trends and Emerging Practices* is broadening so risks are understood and engaged in a wider context. In the past, most projects have only considered the risks of uncertain future events.

Examples of future event-based uncertainty risks include a key seller going out of business during the project, the customer requiring changes after the design is complete, or a subcontractor proposing enhancements to the standard operating processes.

The increasing recognition that non-event risks need to be identified and managed is also covered. Emerging practices focus on two main types of non-event risks. They are:

- *Variability risks* due to uncertainty about key features, activities or decisions. Variability risks include delivery above or below targets, a higher or lower quantity of errors, or something as industry-specific as an unseasonal weather event during construction.

- *Ambiguity risks* due to uncertainty about the future where incomplete information could affect achievement of project objectives. Ambiguity risks include requirements or facets of the technical solution that are unclear, unknown or unanticipated regulatory changes, or a high degree of systemic complexity.

Such risks are managed by framing areas where a deficit of knowledge or understanding exists, then using expert judgement or benchmarking or best practices to find options for acquiring needed information through incremental development, mock ups, or computer simulation. In Agile methods it is common to refer to this type of work as a "Spike."

Project resilience is identified as another emerging trend for dealing with "unknowable unknowns" that are only recognized after they have occurred. This requires the project to have:

- The right level of budget and schedule contingency.

- Flexible project processes for maintaining overall direction.

- Project teams empowered by clear objectives where "done" has been defined within agreed-upon limits.

- A process for frequently reviewing early warning signs.

- Clear input from stakeholders identifying when the project scope or strategy should be adjusted to respond to the appearance of unknowable unknowns.

Integrated risk management requires that project risks be aligned with the organizational context and that they be owned and managed at the appropriate level. The appropriate level can be within the project domain or external to it.

The *Tailoring Considerations* section discusses four considerations:

- *Project size* because size matters. If the project is small, keep it simple. If it is large, due diligence is mandatory.

- *Project complexity* sometimes created by factors like a demand for significant innovation, a reliance on new technology, or significant external dependencies.

- *Project importance* driven by a strategic need, a political opportunity, or an existential threat.

- *Development approach,* especially if a custom, tailored, hybrid approach is being used for the first time.

Considerations for Agile and Adaptive Environments states that projects with high-variability incur more uncertainty and risk. To manage this, adaptive approaches make use of frequent reviews of incremental work products and cross-functional project teams to accelerate knowledge sharing and ensure the risk is understood and managed.

CHAPTER 12

Project Procurement Management

Under *Key Concepts* a number of risks are noted, including:

- Legalities, and the always lurking fine print in contracts.

- Dependencies outside the project that could impact production of key materials or the logistics of customer-supplied items.

- The need for long-lead items, some of which might have to be ordered before product designs and project scope are finalized.

To succeed PMs must fully understand statutory and regulatory requirements, organizational policies and governance procedures, and then wisely cultivate contractor/supplier relationships to align with decisions about contracts.

In *Trends and Emerging Practices* the spotlight is on six key elements:

- *Better tools,* such as the Building Information Model (BIM) used in the architecture, engineering and construction (AEC) industry to facilitate better design practices.

- *Advanced risk management,* where contracts allocate risks to the buyer or contractor based on who has control over it.

- *Changing contract processes,* especially in mega-projects for highly-engineered infrastructure deliverables where the buyer works together with all contractors to create combined procurements that leverage economies of scale in purchases and logistics.

- *Logistics and supply-chain management,* to deal with the challenges of long-lead timeframes by using known requirements defined in the top-level design. Incidentally, that could be considered one purpose of an Agile Roadmap.

- *Technology and stakeholder relations,* especially for publicly funded projects, where constituent expectations have been radically changed due to Google, Amazon and Facebook, requiring the integration of online, on-demand access to all or specific parts of the program information and current status.

- *Trial engagements,* so that the buyer can engage in several contractor candidates for initial deliverables and work products on a competitive, paid basis, then use that experience to accelerate the evaluation of potential partners. This supports the formation of long-term relationships while simultaneously making progress on project work.

The section on *Tailoring Considerations* highlights four elements:

- *Complexity of procurement,* such as using multiple vendors, not a single supplier.

- *Physical location,* where the impact of time zones, logistics and supplies chains vary.

- *Governance and regulatory environment,* where international, national, state and local factors can all impact the procurement.

- *Contractor availability,* because too many and too few each introduce different challenges.

In the section on *Considerations for Agile and Adaptive Environments* the discussion of Agile environments delves into how suppliers may extend the team or be a collaborative relationship to share risk via something like a master services agreement (MSA). An MSA can simplify procurement management by separating the adaptive work into an appendix or supplement so that scope adaptations don't impact management of the overall project contract.

CHAPTER 13
Project Stakeholder Management

Under *Key Concepts* Stakeholders are broadly defined as everyone who is impacted by or can impact, positively or negatively, project success – both real and perceived. Both experience and research show the importance of engaging and managing stakeholders by design, not by accident. It is well-documented that consistent, iterative engagement and re-engagement of stakeholders across all phases of the life-cycle must be integrated with easily-accessible, transparent information aligned to their evolving priorities.

Trends and Emerging Practices notes that broader definitions of stakeholders are developing, and being accepted, extending beyond the traditional categories of employees, suppliers, and shareholders, to include groups such as regulators, lobby groups, environmentalists, financial organizations, the media, and those who simply believe they are stakeholders.

Trends and emerging practices include:

- Rigorous identification of all stakeholders
- Team involved in engagement activities
- Stakeholder consultations for co-creation workouts

- Sharing the benefits derived from active stakeholder support and the true cost of not engaging them

The *Tailoring Considerations* section focuses on three elements:

- *Stakeholder diversity,* suggesting that the number of stakeholders and their cultural diversity plays a significant role in how they can be integrated.

- *Complexity of stakeholder relationships,* which is affected by the sheer number of relationships (i.e., the formula for the number of communications paths is $(n*(n-1))/2))$ in a non-linear way.

- *Communication technology,* which must cover when, where, what, why and how from the stakeholders' perspective and still be cost-sensible.

In the section on *Considerations for Agile and Adaptive Environments* the discussion centers on the need for adaptive teams to engage stakeholders directly to facilitate timely, productive discussions and decision-making. The dynamic co-creative process leads to:

- More stakeholder involvement and higher satisfaction.

- Increased trust and mitigated risk.

- Support for earlier adjustments that reduce costs and increase likelihood of project success.

Agile embraces aggressive transparency to manage the changing project context and its impact on stakeholders.

PART TWO:

THE AGILE PRACTICE GUIDE

Overview

The *PMBOK® Guide* is well positioned to energize the development and implementation of robust, scalable Agile practices. It is much more than just generally accepted project management principles and practices. It has spent decades drawing upon the proven standards and best practices from Generally Accepted Accounting Principles (GAAP), the American Production and Inventory Control Society (APICS) body of knowledge, Lean principles, and general management principles and practices from business schools and universities (Figure 22).

It has now begun integrating leading Agile ideas because its mission and purpose is to be an authoritative curator of proven practices and information used in the project management profession. It is an ANSI Standard with the respect that is due such references because of the rigor applied to its curation process and the contributions invested by many hundreds of professional practitioners. It carries that responsibility with sober discipline and has earned immense marketplace credibility because of it.

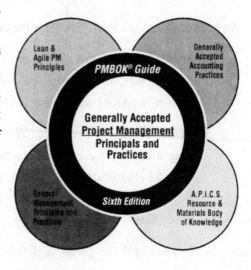

Figure 22

The fact that the *PMBOK® Guide* is **the standard** is very good news for those who want to learn and leverage all the benefits of Agile and are not dogmatically committed to the premise of ending PMI and eliminating project managers as the path to "changing the world of work." Lean and Agile are separate disciplines and independent schools of thought from the *PMBOK® Guide,* as they should be. Integrating the learning and discovery of best practices and proven processes they develop into the *PMBOK® Guide* is complementary and a compliment!

The mission of PMI is authoritative curation while the mission of the Lean and Agile schools is thought leadership. Those missions are complementary. The objective of the *PMBOK® Guide* is to be an outstanding American National Standards Institute (ANSI) Standard and the inclusion of Lean and Agile elements is a compliment to the thought leadership those disciplines and schools of thought are generating.

One benefit to the stakeholders of all three communities – PMI, Agile and Lean – is that the rigorous researching, testing and editing process that precedes information being included in the *PMBOK® Guide* will improve cross-industry communication by standardizing vocabulary.

An example from the Sixth Edition is the defining of life cycles, where it details:

- *Predictive* life cycles set the project scope, time, and cost early in the life cycle, and the manage changes carefully. Predictive is also known as (AKA) waterfall or traditional life cycles.

- *Iterative* life cycles usually set the project scope early, also, but time and cost are expected to be modified as the project team gains a better understanding of the product. The product develops through repeated cycles, iterations, delivering increments that add functionality to the product.

- *Incremental* life cycles add functional deliverables in iterations within an agreed upon timeframe, but contains the complete, sufficient capabilities only after the final iteration.

- *Adaptive* life cycles start with a detailed scope and use iterations to develop the product and can be iterative or incremental. AKA Agile or change-driven life cycles.

- *Hybrid* life cycles combine predictive and adaptive, or multiple adaptive frameworks, to create a life cycle tailored to the needs of the team and environment.

We will now move forward with Deep Dive coverage of the Agile Practice Guide and conclude with an Assessment and some future predictions.

SECTION 1

Introduction to the Agile Practice Guide (APG)

In the introduction the core writing team explains its mindset by saying, "This practice guide provides practical guidance geared for project leaders and team members adapting to an agile approach in planning and executing projects... [with] a practical approach to project agility. This practice guide represents a bridge to understanding the pathway *from* a predictive approach *to* an agile approach." *(Emphasis added.)*

Perhaps it is because we are optimistic skeptics (but not cynics!) and take everything with a grain of salt. Or perhaps it is because we let evidence and and research drive our opinions. Who knows? Maybe it is because we are long-time, serious project management practitioners who are tired of feeling that many Agilists maintain an ill-informed, negative attitude towards PMs with their minds held tightly shut against evidence to the contrary. Whatever the case may be, that opening statement made us flash back to reading the children's story about "the Emperor with no clothes." (Figure 23.)

Figure 23

67

The opening statement seems to belie a predisposition to assume that everyone wants to, or must, move from a predictive approach to an Agile one. It appears to support the hypothesis that every project, everywhere, every time needs to be Agile. But any experienced project manager can tell you, there is not one best way to do every project. It sets up a false dichotomy by presenting an "either or" choice as opposed to a "both and" choice, despite abundant evidence that "market share" growth in hybrid (i.e., "both and") approaches is expanding faster than in any other framework.

The APG continued, that the writing style was more relaxed and informal than is typical for PMI standards. They added that they would incorporate new elements including Tips, Sidebars, and Case Studies to better illustrate key points and concepts. Therefore, we felt it made sense to catalog and score those elements to see how much value they provided as part of the APG. We scored each item using a scale of 1 to 5, with 5 being very high value. There were sixteen "Tips" with an average score of 2.63, fifteen "Sidebars" with an average score of 2.73, and three "Case Studies" with an average score of 1.

In the last paragraph of page one, they also explained that the APG would go beyond addressing just Agile in software development and expand into non-software environments. Again, we felt that provided a useful, quantifiable metric to evaluate how much useful information the APG provides in non-software environments. Sadly, the data shows 16 Tips with 1 being non-software (6.5%), 15 Sidebars with 2 qualifying (13%) and 3 Case Studies with none (except for two that referenced the writing process used by the team for the APG, which just wasn't useful or credible).

On page 2, they ask the question, "Why now?" and then vaguely acknowledge that project teams have been using Agile techniques for "at least several decades" before the arrival of the *Agile Manifesto*. They also note, that "... in order to stay competitive and relevant, organizations can no longer be internally focused but rather need to focus outwardly to the customer experience." (Figure 24.)

Figure 24

We would agree that the playing field for both public entities and commercial industries is one of rapid, disruptive change! Agile can definitely add value, just not on *every* project and not as an "either or" choice for the whole enterprise.

On page three it says, "The Agile Practice Guide is project focused and addresses project lifecycle selection, implementing agile, and organizational considerations for agile projects. Organizational Change Management (OCM) is essential for implementing or transforming practices but, since OCM is a disciplined within itself, it is outside the scope of this practice guide."

We find this statement to be something of an oxymoron. First, it says that it is focused on implementing Agile and organizational considerations. Then it says organizational change management is essential. It finishes by saying organizational change management is outside of the scope of this practice guide. We find this to be similar to saying the guide is focused on teams needing servant leaders, however servant leadership is a separate discipline and is out of scope.

On page 4, Table 1-1 provides a more in-depth list of things that are in scope and items that are out of scope. It has a couple of uses. First, for purposes of PMI-ACP® exam takers, it is a good idea to assume the *in-scope* items will be included in the exam. Second, it provides a useful outline of the stated mission of the APG, allowing you to assess if it is too shallow, delivers as a minimum viable product (MVP), or exceeds an MVP and is useful or robust in addressing your needs and interests.

Things considered in scope:

1. Implementing Agile at the project or team level.

2. Popular Agile approaches, as listed in industry surveys.

3. Suitability factors for choosing an Agile approach and/or practice.

4. Mapping Agile to the *PMBOK® Guide.*

5. How to use Agile outside of software development.

6. Guidance for implementing Agile in the real world.

7. Definitions of generally accepted terms.

Things considered out of scope:

1. Implementing Agile programs or organizational (scaled) Agile.

2. Coverage of industry-specific methods or tailored life-cycle techniques.

3. Advocating particular methods or approaches.

4. Suggesting *PMBOK® Guide* improvements.

5. Mitigating software industry bias on Agile approaches.

6. Step-by-step guidance for implementing Agile.

7. Defining new terms.

You can evaluate if the things defined as in scope are adequately covered and whether things left out of scope should have been included, and then create a scorecard. Our scorecards are shown below.

The In-Scope Scorecard (Figure 25) assessment found that the coverage of Agile at the project and team levels was MVP and maybe even usable, but it did not add anything to Agile content easily available from other sources where it is more clearly expressed. Discussion of popular Agile approaches was equivalently assessed.

Choosing an Agile approach and the attempt to map Agile to the *PMBOK® Guide* did not rise to the level of being an MVP, although there were instances where it was nearly there.

IN SCOPE SCORECARD	SHALLOW	MVP	USABLE	ROBUST
Description				
1 Agile at Project & Team Levels		◆	◆	
2 Popular Agile Approaches		◆	◆	
3 Choosing an Agile Approach	◆	◆		
4 Map to the *PMBOK® Guide*	◆	◆		
5 Agile Outside of Software	◆			
6 Guidance Implementing Agile	◆			
7 Define Generally Accepted Terms		◆	◆	

Figure 25

The treatment of Agile outside of software environments was virtually absent and the guidance regarding implementation considerations was completely inadequate. Lastly, the definitions provided rose to the level of MVP and flirted with being valuable, but did not add anything to content easily available and more clearly expressed elsewhere.

The Out-of-Scope Scorecard (Figure 26) assessment was that the coverage of Agile for Programs really was, perhaps, required, or at a minimum needed, because that is the state of the marketplace and there is a shortage of vendor-agnostic, pragmatic content available despite a clear need. It represented a high-value opportunity that was missed. Discussion of industry-specific approaches would definitely be helpful, but was a purely optional choice for an industry-wide guide.

When it came to Advocating Particular Approaches, offering vendor-specific suggestions would have been too biased for an industry-wide guide, but comparing the strengths and weaknesses of particular approaches could have filled a need and certainly would have been helpful.

Because of the co-release timing, outlining important areas for *PMBOK® Guide* Improvements in the future would definitely qualify as helpful to the practitioner community, but was a purely optional choice. In comparison, the need to Mitigate Software Industry Bias was clear and present, and addressing it would have been extraordinarily helpful.

OUT OF SCOPE SCORECARD	REQUIRED	NEEDED	HELPFUL	OPTIONAL
Description				
1 Agile for Programs	♦	♦		
2 Industry-Specific Approaches			♦	♦
3 Advocating Particular Approaches		♦	♦	
4 *PMBOK® Guide* Improvements			♦	♦
5 Mitigate Software Industry Bias		♦	♦	
6 How-To Implementation Guidance		♦	♦	
7 Defining New Terms		♦	♦	

Figure 26

Defining How-To Implementation Guidance as out-of-scope seemed odd since Guidance Implementing Agile was defined as in-scope previously. The differences between the two were more nuanced than the short titles suggest, but since it is an area of great interest to practitioners, it was another opportunity missed. Lastly, Defining New Terms caused some real consternation because they did, in fact, define several terms in new ways, including Dispersed Teams and Distributed Teams on page 43 only to misuse the terms on page 44 and beyond. (As fellow authors, we sympathize with how difficult it is to catch all the mistakes during editing so this observation is not a condemnation or negative criticism.)

The vital point is that each practitioner needs to exercise care while assessing when and how much of the APG to use for their teams, projects and organization.

The APG is organized as follows:

SECTION 1 – Introduction (to the Agile Practice Guide) (6 pages)

SECTION 2 – An Introduction to Agile (10 pages)

SECTION 3 – Lifecycle Selection (15 pages)

SECTION 4 – Implementing Agile: Creating an Agile Environment (16 pages)

SECTION 5 – Implementing Agile: Delivering in an Agile Environment (21 pages)

SECTION 6 – Organizational Considerations for Project Agility (16 pages)

SECTION 2
An Introduction to Agile

Sadly, Section 2 feels like an opportunity missed because it is a rather shallow review of information that is commonly available through any number of other sources with no real value-added.

It starts with a discussion of the differences between definable and high-uncertainty work, using industries where it is presumed that there is high-uncertainty during design but definable, predictable work during production (Figure 27). But the examples of predictable situations fly in the face of the Lean principles that originated in the Toyota Production System (TPS) precisely because of the complexity involved in production. A case study of this erroneous suggestion can be seen with even a cursory study of Tesla's experience with ramping up production volume for car assembly.

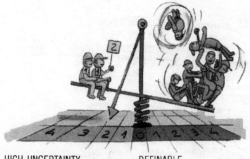

HIGH-UNCERTAINTY
• Discovery & Learning Required
• New Field or Operating Environment

DEFINABLE
• Clear Procedures
• History of Similar Projects

Figure 27

On page 10, Figure 2-3 presents an illustration that may cause a lack of credibility in the minds of serious, seasoned, project management professionals and senior organizational leaders. It appears to suggest that an Agile Mindset is the progenitor of values, which beget principles and practices, implying that somehow one must become "enlightened" in order to use Agile and enjoy its benefits. The persistent dogmatic mantra, "Mindset, mindset, mindset!" has limited the appeal of Agile to the "Micro-Dynamic" arena of small teams, small projects, and small initiatives. Appearing to allude to this quandary, just under the illustration, it says, "… The approaches and techniques being used by project teams today existed before the agile manifesto by many years and, in some cases, decades." The logical disconnect becomes clear in the very next illustration, Figure 2-4, which shows that Lean principles are the foundation, or progenitor, of Kanban and Agile.

A more accurate, and perhaps more useful, analogy would be a Banyan tree where knowledge grows from Galileo and the scientific method, to insights from Shewhart, TPS, and Deming expressed in the Plan, Do, Check, Act (PDCA) cycle. From there we see Kanban as a connector between the *PMBOK® Guide* and *Agile Manifesto* surrounded by various other disciplines and tools. The tree put down growth on the *Manifesto* side such as Scrum, XP, and Lean Software Development, sharing a common focus on the team-level dynamic on one-side, and growth on the *PMBOK® Guide* side such as Disciplined Agile and Scaled Agile Framework (SAFe®), sharing a common focus on the program- and enterprise-level dynamic (Figure 28).

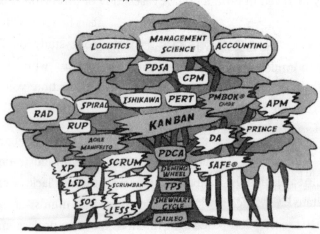

Figure 28

Before going any further, we need to explain that there are times we tell our clients and students that the "real word" and the world of the *PMBOK Guide°* – or in this case the *Agile Practice Guide* – don't align and it becomes necessary, ***especially for exam takers,*** to don the proverbial "tinfoil helmet" (to prevent anyone from attacking your brain). There are times when Agile-speak or Agile dogma induces the need for a tinfoil helmet also. When those occasions arise, we will use the "ALERT – TINFOIL HELMET REQUIRED" icon, shown here, to draw your attention to them.

The first example of a Tinfoil Helmet alert occurs on page 10 where it says, *"**Agile approaches** and **agile methods** are umbrella* terms that cover a variety of frameworks and methods." It italicizes the terms Agile approaches and Agile methods, implying that those two terms have significance, potentially, for exam takers. However, neither of those terms is found in the glossary!

 A second instance occurs in the sidebar on the same page, where it says, "Is agile an approach, a method, a practice, a technique, or a framework? Any or all of these terms could apply depending on the situation. This practice guide, uses the term "approach" unless one of the other terms is obviously more correct." It fails, however, to explain how to differentiate when one term versus another is more obviously more correct.

On page 12, it says, "In general, there are two strategies to fulfill agile values and principles. The first is to adopt the formal agile approach... The second strategy is to implement changes to... progress on a core value or principle." We would add that the first strategy is most often applied when Scrum is used to introduce Agile and pilot it with a team. The second strategy is more often used when an organization begins scaling Agile.

In Section 2.3 Lean and the Kanban Method, the wording is very stilted. What it is trying to say is that the Kanban Method that emerged in the mid-2000s did so to facilitate knowledge work by offering a continuous flow approach as an alternative to the iteration-based agile methods that were prevalent at the time.

SECTION 3
Life Cycle Selection

The APG refers to four types of life cycles, named as follows:

- Predictive life cycle (i.e., Traditional)

- Iterative life cycle

- Incremental life cycle

- Agile life cycle

For *exam takers*, it is good to note that the APG defines each of the life cycles as follows:

> *Predictive life cycles* work in environments with reduced uncertainty and complexity where project work is known and development approaches proven so that teams can segment work into sequential predictable batches.

> *Iterative life cycles* use feedback on unfinished work in order to adapt expectations about the remaining work and improve the development approach used.

> *Incremental life cycles* deliver finished units of work that are part of the planned final result but not all of it, because the customer

expects to be able to use them immediately and, in doing so, get benefits that contribute to a better financial return on investment.

Agile life cycles integrate both iterative and incremental characteristics to create customer visibility sooner then use that early feedback to deliver the highest value finished deliverables. Doing so builds customer confidence that the team's approach can potentially lead to an earlier release accelerating return on investment and project success.

During a deep discussion on this section with Amazon #1 Best Selling author, Doug Martin, PMP, PMI-ACP, CSM, CSP, he made the point that this is an area where the APG seems to fail to recognize, and certainly fails to clarify, the difference between the method used to develop a product and the release schedule for that product. Iterative and Incremental life cycles are both development and release models. While it can be argued that normally they align, there are situations where development may happen iteratively but the release occurs incrementally. Consider, as an example, a Nuclear Power Plant Control System where Release 1.0 has to be relatively robust, exceeding the common standard of an MVP, and the development work to deliver Release 1.0 occurs incrementally with subsystems like Safety Control and Power Generation systems, which must be continuously tested as subsequent layers are added. The example can be extended by envisioning Release 2.0, which needs to be released as a packaged solution because of the consumer market stakeholders who will use it, but, again, development incrementally fulfills the product release schedule.

Another way to visualize the relationship between Iterative, Incremental and Agile approaches is as expanding layers of value delivery. At the core, Iterative approaches allow for exploration as an unavoidable necessity in high-complexity environments and deliver value through learning and discovery (Figure 29).

Incremental methods use learning and discovery and extend it to build subcomponents that can be used as prototypes to help the customer better understand and articulate their needs with feedback that leads to "actionable insight."

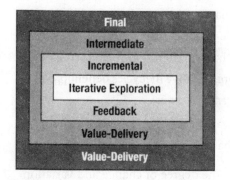

Figure 29

Agile methods integrate exploration activities and subcomponents with other potentially usable units of progress to deliver intermediate value, and then, final value.

As the size of the vision or objectives or requirements increases it may become beneficial to integrate predictive methods to guide planning at the strategic or tactical level, referred to as ROM or Affinity estimating. Planning at that higher Program, Release or Roadmap level means lower resolution of the requirements and lower granularity of the time scale, reducing uncertainty and complexity by definition. The result is estimates require less precision, and predictive forecasts become possible.

In the Sidebar on page 20, "Planning Is Always There", the APG points out, "A key thing to remember… is not whether planning is done, but rather how much planning is done and when. At the predictive end of the continuum, the plan drives the work… In iterative approaches, prototypes and proofs… are intended to modify the plans… and inform future project work. Meanwhile, incremental initiatives plan to deliver successive subsets… [and] plan several successive deliveries in advance or only one at the time… [to] inform the future project work. Agile projects also plan… the team plans and re-plans… [using learning and discovery] from reviews of frequent deliveries."

On page 20, it also says, "Predictive life cycles expect to take advantage of high certainty around firm requirements… [so] the team requires detailed plans to know what to deliver and how… [and those] detailed requirements and plans… articulate the constraints… to manage risk and cost… predictive projects do not typically deliver business value until the end of the project…[so] if changes or disagreements [are encountered] the project will incur unanticipated costs."

79

Exam takers should note, Section 3.1.2 Characteristics of Iterative Life Cycles defines them as those that, "...improve the product or result through successive prototypes or proofs of concept." In that way, iterations help identify and reduce uncertainty in the project. Therefore, iterative lifecycles benefit projects facing high-complexity, utilizing learning and discovery.

Exam takers should also note, Section 3.1.3 Characteristics of Incremental Life Cycles defines them as, "... optimized for speed of (project) delivery." The frequent delivery of smaller outcomes means the degree of change and variation is less important than ensuring customers get value sooner than the end of the project.

Exam takers need to be aware Section 3.1.4 Characteristics of Agile Life Cycles defines them as environments where "...the team expects requirements to change." The purpose of using both iterative and incremental approaches is to uncover hidden or misunderstood requirements based on customer feedback, then better plan the next project phase.

It also emphasizes that, "Agile life cycles are defined as those that fulfill the principles of the Agile Manifesto." It is unclear if that is a reference to the specific *12 Principles Behind the Agile Manifesto* published by the co-authors of the *Agile Manifesto* or a broader reference to other principles.

Exam takers need to remember Section 3.1.4 Characteristics of Hybrid Life Cycles defines them as combining different elements of life cycles with the intention of better achieving the project objectives. Those combinations may include a blend of predictive practices with iterative, incremental, and Agile methods, or a blend of only iterative, incremental and Agile methods, but both constitute a hybrid approach.

On page 30, 3.1.11, Hybrid Life Cycles As Transition Strategy describes a two-step approach to changing. Step one involves improving learning and alignment using new (Agile) techniques on less risky projects. Step two accelerates value and return on investment by applying the methods to more complex projects. It points out that making the hybrid transition is dependent on two variables; the organization's situation and the team's readiness to adapt and embrace change. The viewpoint seems to be "either/or" thinking. Either the organization is Agile or it is Traditional. In practice, however, hybrid is often used as a "both/and" approach that integrates Agile without abandoning Traditional because of the recognition that they key to success is aligning the methods with the unique needs of the project, program or portfolio.

The guide also advises, in Section 3.2 Mixing Agile Practices, that, "Agile frameworks are not customized for the team. The team may need to tailor practices to deliver value on a regular basis. Often, teams practice their own special blend of agile, even if they use a particular framework as a starting point." Contrasting the first point, frameworks are not customized, with the second point, teams use their own special blends, presents an artificial distinction where there is no real difference. It also seems to support the prior point about either/or versus both/and paradigms. Mixing practices is hybrid, by definition, and hybrid is a both/and approach to Agile.

Implementing: Creating an Agile Environment

This section opens with the statement, "Managing a project using an agile approach *requires* the project team to adopt an agile mindset." *(Emphasis added.)* Yet a few lines farther down, it says, "When a cross-functional team delivers finished value often and reflects on the product and process, the teams are agile. It doesn't matter what the team calls its process." These two statements seem to be in conflict, and the first one seems to demonstrate a common bias amongst Agile aficionados, where an exaggerated emphasis on "mindset, mindset, mindset" comes across like religious dogma. While we don't dispute that adopting or adapting to a people-centric mindset may be difficult for low-skilled project management practitioners, experience has shown that highly skilled practitioners always have a people-centric perspective.

In section 4.2, Servant Leadership Empowers The Team, along with the common content readily available elsewhere, it says, "The servant leader has the ability to change or remove organizational impediments to support delivery teams." It is unclear if they mean that a person who is unable to change or remove organizational impediments is not, or cannot be, a servant leader. In practice, some of the best servant leaders we've ever worked with have been in organizations that did not have a culture that fostered servant leaders, yet their ability to guide, coach and mentor the team made their presence all the more important and indispensable.

Section 4.3 Team Composition opens with a grey-box Tip on page 38 that says, "Build projects around motivated individuals. Give them the environment and support they need and trust them to get the job done." That statement provides a number of key insights. The first sentence, of the two, carries with it the assumption that practitioners have the ability to form teams using only "motivated individuals" even though that is often not the case in the real world where team members are assigned without regard to their motivational state. Otherwise it demonstrates the APG authors' unconscious bias that because they are Agile and motivated, everyone else who comes into contact will become similarly motivated and Agile. The second sentence creates an impression that is ripe for misunderstanding unless we assume they mean "trust but verify" when they say trust the team. The reciprocity of a trust-centric relationship is that it is validated by transparent, honest communication of progress so that stakeholders don't feel like they are being asked to place "blind trust" in the team.

ALERT

TINFOIL HELMET
REQUIRED

Exam takers should note, Section 4.3.1 Agile Teams says, "In practice, the most effective agile teams tend to range in size from 3 to 9 members. Ideally, agile teams are colocated in the team space. Team members are 100% dedicated to the teams." The part of the statement that sizes effective Agile teams from 3 to 9 members varies from the more common guideline of 7 plus or minus 2 taught in many Scrum and Agile classes. But the statement about colocation is good news for exam takers. It means that exam takers can assume the simplest of all environments unless the stem of the question on the exam states otherwise. Of course, that means exam takers may need a "tinfoil helmet" in order to keep from falling into answers based on the real world where team members seldom have the luxury of not being fractionalized across multiple projects and get to be colocated and 100% dedicated to a single team and project.

Exam takers should also take careful note, Section 4.3.2 Agile Roles identifies three common roles and adds a name not yet in common use. They are:

- Cross-functional team members

- Product owner

- Team facilitator (not yet in common use)

It goes on to define each as follows:

> *Cross-functional team members* have all the skills necessary to produce a working product. They are professionals who deliver potentially releasable products on a regular cadence.

> *Product owners* are responsible for guiding the direction of the product. They rank work based on its business value, and work with their teams daily, providing feedback and setting direction regarding the next piece of functionality to be developed and delivered. Further, the product owner works with stakeholders, customers, and the teams to define the product direction.

> It also indicates that Product owners need training on how to organize and manage the flow of work through the team, apparently assuming this responsibility from the project manager at the higher level or the team itself at the lower level. The APG notes one critical success factor for Agile teams is strong product ownership.

> *Team facilitators* are servant leaders, and may go by the names project manager, scrum master, project team lead, team coach, or team facilitator.

Figure 30 visually compares the lexicons of various Agile frameworks.

Two things Exam takers should note occur on page 42, in Section 4.3.3 Generalizing Specialists, which includes a Sidebar about "I-shaped and T-shaped People."

The APG authors attempt to make the case that team members should possess a broad skill base, be T-shaped, rather than having a narrow specialization,

APG & AGILE'S "BIG 5"	APG	SCRUM	XP	LSD	KANBAN	HYBRID
ROLES						
Customer	■					
Cross-functional Team	■					
Product Owner	■					
Team Facilitator	■					
Scrum Master		■				
Architect, Coach, Programmer			■			
Product & Project Managers						
Tracker						
Interaction Designer & Batman			■			
Tester			■			
Analysts, Developers, Support						

Figure 30

I-shaped. However, real world experience is more nuanced. The team needs a broad skill set, which is why it must be cross-functional, but that does not mean that each team member cannot be a specialist. While the concept of "Generalizing Specialists" has become religious Agile dogma, and even though the hypothesis that the type of person able to master multiple skills has a strong ability to learn new ones quickly sounds good, it has fatal flaws for real-world teams.

As explained above in Part One on the *PMBOK® Guide – Sixth Edition* the Generalizing Specialists approach has a nearly-fatal flaw because it assumes the availability of specialists who are zebra-striped giraffes when Agile offers a much more reliable solution with a systemic approach that creates easy-to-hold batons of tribal knowledge.

Figure 31

As Doug Martin pointed out, this is indeed a struggle in the real world of software development. If a team was lucky enough to have a resource that was truly dedicated to just one project, and that resource's primary value was a narrow, specific skill set, the odds of always having an iteration backlog right-sized to the team and also being able to keep that specialized resource busy full time is extremely low. That means choosing to fractionalize the resource across more projects, an admittedly undesirable choice, or having them perform other tasks outside of their specialty. The real-world problem is that the specialized resource needs training, standards and checklists to be effective in performing other tasks, entry-level activities, but those checklists and instructions don't exist. Every team member is bright and can follow a recipe, which helps them develop knowledge and abilities, but without the recipe the assumption of a "cross-training fairy that swoops in and telepathically educates them" becomes necessary.

Exam takers please notice that Section 4.3.4 Team Structures includes two terms that are not in the Glossary. From the content on pages 43 and 44 it appears the definitions are:

ALERT

TINFOIL HELMET
REQUIRED

Distributed teams have *teams* in various locations.

Dispersed teams have *members* in various locations.

The reason it only *appears* the definitions described above are accurate is because on page 46, in Section 4.3.6 Team Workspaces, in the third paragraph, it says, "When teams have geographically distributed members..." which conflicts with the terms distributed and dispersed as previously defined on page 43. It should say, "when teams have geographically dispersed members..." The confusion, even by the APG authors, should serve as a caution to exam takers to be careful how they use those two terms.

At the bottom of the same page they suggest considering *fishbowl windows* and *remote pairing* for managing communication with dispersed teams. Again, the confusion of terms occurs, because it is either dispersed members or distributed teams. In addition, while the two terms fishbowl windows and remote pairing are italicized for emphasis and defined with bullet points, there is no entry for those two terms in the glossary.

Fishbowl window is defined as "setting up long-lived video conferencing links between the various locations." Such linkage allows team members to see and engage spontaneously with each other because the video connection stays live all day, increasing total collaboration for the team.

Remote pairing is defined as "using virtual conferencing tools to share screens, including voice and video links." Again, this helps increase team collaboration, but is limited to situations where time zone differences aren't too large.

Both tools increase the growth of tribal knowledge.

Implementing: Delivering in an Agile Environment

5.1 Charter the Project and the Team

This section provides one of our favorite insights. The idea of chartering not only the project but also the team is an excellent one. Project charters help the team know "why this project matters" and what this project's objective is, creating an opportunity for a Team charter that aligns the standards for working agreements to best support those objectives and that strategy.

Agile teams work in close proximity – sometimes physical, sometimes not – but always close creative and development proximity, which is why they require team norms, often called Working Agreements, to unify around an understanding of how they will work together. Expressing that in a team charter is a natural and excellent application of that principle to meet the need for a mutually supportive environment of personal safety.

The APG suggests Agile charters answer these questions:

- Why are we doing this project? (Project vision)
- Who benefits and how? (Project vision and / or purpose)
- What does "done" mean for the project? (Project release criteria for the team)

- How are we going to work together? (Intended flow of work for the team)

The first two bullets seem project-centric and the second two seem team-centric.

ALERT

TINFOIL HELMET
REQUIRED

Exam takers might want to note that on page 50 the APG says, "Teams do not need a formal process for chartering as long as the teams understand how to work together." *(Emphasis added.)* This statement seems to be in direct conflict with the statement that opened this section where it says, "Agile teams *require* team norms and an understanding of how to work together." *(Emphasis added.)* The truth is, in most environments, most teams will not work well together without some process for creating a mutually-accepted and understood team dynamic. Suggesting they do not need a process – unless the sentence expressed a bias or aversion of the authors for anything *formal* – makes no sense, and even then, it makes no sense!

Team charters create a "social contract" that helps the team move more effectively and efficiently through the Forming and Storming stages of Tuckman's model, and the APG provides some ideas to use as a basis for creating the charter. The ideas include:

- Sustainable pace and other team values
- Core hours and other team practices
- Working standards like the definition of done
- Ground rules like attack the problem not the person
- Group norms like being on time for the daily meeting

Each element helps clarify the team charter by expressing how the members will interact with each other.

In Section 5.2, Common Agile Practices, a number of practices are named and discussed. They include:

- Retrospectives, where no unusual, novel or new content is introduced.

- Backlog preparation, where the idea of a Product Owner Value Team is introduced and seems to relate to larger scaled Agile.

- Backlog refinement, where the guide states that there is no consensus on how long the refinement should take because it is a continuum. It also states that the product owner conducts backlog preparation and refinement meetings. It also endorses the use of the word "refinement" instead of the older word "grooming," which has occurred due to the undesirable character that the word grooming has acquired in reference to crimes against children.

- Daily Standups, where no new content is introduced.

- Demonstrations/Reviews, where they state, "The product owner sees the demonstration and accepts or declines the stories."

ALERT

Exam takers and Agile practitioners might find this statement about Demonstrations and Reviews a bit of a surprise because it seems to violate the idea of working closely with the "voice of the customer." In many real-world environments, the product owner is regularly involved with the development team, guiding the development team's understanding by approving user stories as they are completed so that at demo and review meetings the product owner and team stand together as the increment is presented to the various stakeholders and receive their feedback jointly.

- Planning for Iteration-Based Agile, where no new content is introduced, but a reference is made to see Section 4.10 for examples, but Section 4.10 does not exist.

- Execution practices that help teams deliver value, which includes some bullets for:

  - Continuous integration

  - Test at all levels

  - Acceptance Test-Driven Development (ATDD)

91

- Test-Driven Development (TDD) and Behavior-Driven Development (BDD)
- Spikes
- How Iterations and Increments Help Deliver Working Product where no new content is introduced.

In Section 5.4, Measurements in Agile Projects, there is, once again, a reference to Section 4.10 which does not exist. It also says, "In addition to the quantitative measures, the team can consider collecting qualitative measures." As it continues, it describes qualitative measures with examples that could probably be more accurately described as subjective measures.

ALERT

TINFOIL HELMET
REQUIRED

Exam takers, especially PMP® credential holders who are exam takers, should note that the APG does not follow the PMBOK® Guide use of the term qualitative to refer to the probability and impact of risk, nor the use of the term quantitative to refer to the cost or time value of risk. The way the APG uses those terms creates possible confusion for those with experience and training aligned with the *PMBOK® Guide* because the use of qualitative and quantitative in this section is not aligned with their prior training.

Section 5.4.1, Agile Teams Measure Results contains examples of charts for Burn-Up and Burn-Down, Features Complete, Remaining and Total, and Product Backlog Burnup, as well as a Kanban Board and a Cumulative Flow Diagram, which all are standard fare. However, on page 66, the APG states, "Burnups, burndowns (capacity measures) and lead time, cycle time (predictability measures) are useful for in-the-moment measurements." By contrasting them as "capacity" and "predictability" measures, the authors provide some interesting food for thought.

ALERT

TINFOIL HELMET
REQUIRED

Exam takers, especially PMP® credential holders who are exam takers, need to note Earned Value has two new variables. Earned Schedule theory has expanded the metrics used in traditional EVM with two new variables, Earned Schedule (ES) and

Actual Time (AT), used in the equations ES – AT and ES / AT, as explained above in Part One on the *PMBOK® Guide – Sixth Edition.*

Organizational Considerations for Project Agility

This section opens by saying, "While project leaders may not have the ability to change organizational dynamics as they see fit, they are expected to navigate those dynamics skillfully." We agree with that assessment and see it as countering the prior statement in section 4.2 where, it said, "The servant leader has the ability to change or remove organizational impediments to support delivery teams." Changing the culture or removing specific obstacles may not always be possible, but skillfully navigating organizational dynamics remains an option.

Section 6.3, Procurement and Contracts, points to the *Agile Manifesto* value statement that says "Customer collaboration over contract negotiation" and suggests several contracting techniques that can support the necessary dynamic for Agile environments. The list includes:

- Multi-tiered structures
- Emphasize value delivered
- Fixed priced increments
- Not to exceed Time &Materials
- Graduated Time &Materials

- Early cancellation option

- Dynamic scope option

- Team augmentation

- Favor full service suppliers

Section 6.5.1, Frameworks, includes an interesting quote that seems to summarize why Agile is limited in many environments. It says, "The guidance of the most widespread agile methods such as Scrum and eXtreme Programming focus on the activities of a single, small, usually colocated, cross-functional team. While this is very useful for efforts that require a single team, it may provide insufficient guidance for initiatives that require the collaboration of multiple agile teams in a program or portfolio."

The remainder of Section 6 provides very cursory coverage of scaling Agile. It is so thin that it provides very little value, and certainly no information that will help practitioners in the real world.

SECTION 7

A Call to Action

Section 7 closes the APG with a few comments and an invitation to join their blog audience.

Help Your PMI Chapter Excel!

You can get *PMI-ACP® certified*...the *Easy Guaranteed Way*... right in your hometown! Classes include amazing *Instructors* and *On-Demand* content so you get the best of *expert interaction* and *self-study convenience*. All backed by a *100% Money-back Guarantee* that you will pass on the *First Try* that is good for a full year after the class!

Your PMI Chapter can partner with GR8PM for PMI-ACP® Exam Prep and provide *remarkable Member value!* We also make it super easy for (overburdened) Chapter leaders and they'll love that everything we do is offered 100% Risk-free for the Chapter. The classes produce income to the Chapter, and help the Chapter launch its own program when they decide the time has come!

Just introduce us via email to your Chapter Leaders and we'll take it from there.

The GR8PM Team (ops@gr8pm.com)

About the Author

JOHN G. STENBECK, PMP, PMI-ACP, CSM, CSP, is the Founder of GR8PM, Inc. (pronounced "Great PM"). John is a three-time *Amazon #1 Best Selling* author. His industry-leading project management books have reached almost $3 million in sales. His next book, *Distilled Insight: Enterprise Agility in Healthcare*, will be released in September, 2018, and is highly anticipated to become an immediate Amazon #1 Best Seller also.

He has been a guest on VoiceAmerica talk radio, Good Morning America, the Today Show, Fox News and the Oprah Winfrey Show. John has been featured on the front page of the Los Angeles Times and the San Diego Union.

John is a sought-after Keynote speaker because he enables executives and professionals to become *powerfully productive leaders!*

A partial list of John's clients includes: Booz Allen Hamilton, Inc., McLean, VA; County of Orange, Orange, CA; Hewlett-Packard Company, Palo Alto, CA; Lucent Technologies, Allentown, PA; Nike Corp., Beaverton, OR; Oracle Corp., Redwood Shores, CA; Qualcomm, Inc., San Diego, CA; U.S. Army – Space and Terrestrial Communications Directorate, Fort Monmouth, NJ; U.S.D.A. – National Finance Center, New Orleans, LA; U.S. Marine Corp. Systems Command, Stafford, VA; Visa – Smart Cards, Foster City, CA.

He recently recorded 71 video lectures for Villanova University's online certificate programs. John has taught numerous public and corporate on-site programs to over 12,000 participants. John helps technical professionals master project management and leadership skills dramatically improving their contribution to enterprise results.

John holds PMI's Project Management Professional (PMP®) credential and Agile Certified Practitioner (PMI-ACP®) certification. He also holds Certified Scrum Master (CSM) and Certified Scrum Professional (CSP) designations from the Scrum Alliance and an ITIL v3 Foundations certification.

About the Production Team

About the Editor

Lauren Seybert Mix, President, LM Editing Co., has a unique background that makes her a gifted editor of technically sophisticated books. She works on selected projects where she feels her background can provide more than just spelling and grammar checks. The manuscripts she edits involve collaborations where her value as a neutral and independent expert adds to the competence of the author, assisting in the struggle to find the clearest, most accessible way to create maximum knowledge transfer for the reader.

In her "spare" time, she enjoys caring for and playing with her 27 animals on her farm in New England.

Lauren's husband, Patrick, and daughter, Mallory, complete her.

Lauren Mix, Owner/President
LM Editing
(615) 815-7021
LMix7712@gmail.com

About the Graphic Designer

Over the last 25 years, Tamara Parsons, owner of KenType, has taken the natural gift of "an eye for graphics" and developed it into an unparalleled level of expertise in graphic design and print production.

A referral brought Tamara and John Stenbeck together for his first book. Finding that they were both driven by a desire to constantly deliver amazing world-class results, an alliance rooted in mutual respect and trust was formed. With Tamara's unique skills both in graphic design, and as an organized multitasking maven, her collaboration with John Stenbeck and his team of co-authors has continued to grow.

John has been quick to refer Tamara to his clients, and his idea – insistence really! – that this tribute be included here. He invites you to contact KenType and experience how much better your results could be!

Kensington Type & Graphics
(619) 281-1520 | tami@kentype.com

About the Illustrator/Artist

We are grateful for the help and support of our illustrator and artist because he makes our technical, dry content and makes it visually rich. He delivers graphic brilliance delivered with elegance, simplicity, passion and creativity.

Please feel free to reach out to him.

Gian Luca Rigliano
Inkolla Graphics (Bologna, Italy)
Facebook Page: inkolla graphic
https://www.flickr.com/photos/artgian-gianluca